YOUR KNOWLEDGE HAS VALUE

Purvesh Shah

Structure and synthesis of Vitamines

GRIN Verlag

Bibliografische Information der Deutschen Nationalbibliothek:

Die Deutsche Bibliothek verzeichnet diese Publikation in der Deutschen National-bibliografie; detaillierte bibliografische Daten sind im Internet über http://dnb.d-nb.de/ abrufbar.

Imprint:

Copyright © 2014 GRIN Verlag GmbH
Druck und Bindung: Books on Demand GmbH, Norderstedt Germany
ISBN: 978-3-656-61353-4

GRIN - Your knowledge has value

Der GRIN Verlag publiziert seit 1998 wissenschaftliche Arbeiten von Studenten, Hochschullehrern und anderen Akademikern als eBook und gedrucktes Buch. Die Verlagswebsite www.grin.com ist die ideale Plattform zur Veröffentlichung von Hausarbeiten, Abschlussarbeiten, wissenschaftlichen Aufsätzen, Dissertationen und Fachbüchern.

Visit us on the internet:

http://www.grin.com/

http://www.facebook.com/grincom

http://www.twitter.com/grin_com

STRUCTURE AND SYNTHESIS OF

VITAMINS

By

Dr. Purvesh J. Shah

Department of Chemistry,

Shree P.M.Patel Institute of P.G.Studies and research in science,

Anand-388001, India

Affilited to Sardar Patel University, Vallabh Vidyanagar 388 120,

India

Contents

Vitamin A

❖ **Another Name: vitamin A_1/ Axerophthol/ Retinol**

❖ **Molecule formula:$C_{20}H_{30}O$**

- Vitamin A_1 influences **growth in animals**, and also apparently increases resistance to disease.

- **Night blindness** is due to vitamin A_1 deficiency in the human diet, and a prolonged deficiency leads to hardening of the cornea, etc.

- Vitamin A_1 **occurs** free and as esters in fats, in fish livers and in blood.

- It was originally **isolated** as viscous yellow oil, but later it was obtained as a crystalline solid, m.p. 63-64°C.

- Vitamin A_1 is **estimated** by the blue colour reaction it gives with a solution of antimony trichloride in chloroform. It is also estimated by light.

- On **catalytic hydrogenation**, vitamin A_1 is converted into perhydrovitamin A_1, $C_{20}H_{40}O$; thus vitamin A_1 contains five double bonds.

- vitamin A_1 forms an ester with p-nitrobenzoic acid, it follows that vitamin A_1 contains a hydroxyl group. Thus the parent hydrocarbon of vitamin A_1 is $C_{20}H_{40}$, and consequently the molecule contains one ring.

- Ozonolysis of vitamin A_1 produces one molecule of geronic acid per molecule of vitamin A_1, and so there must be one *β- ionone* nucleus present.

- Oxidation of vitamin A_1 with permanganate produces acetic acid; this suggests that there are some —$C(CH_3)=$ groups in the chain.

- All of the foregoing facts are in keeping with the suggestion that vitamin A_1 is half the β-carotene structure.

- When heated with an ethanolic solution of hydrogen chloride, vitamin A_1 is converted into some compound (II) which, on dehydrogenation with selenium forms 1: 6-dimethyl naphthalene, III. Heilbron *assumed* I as the structure of vitamin A_1, and explained the course of the reaction as follows:

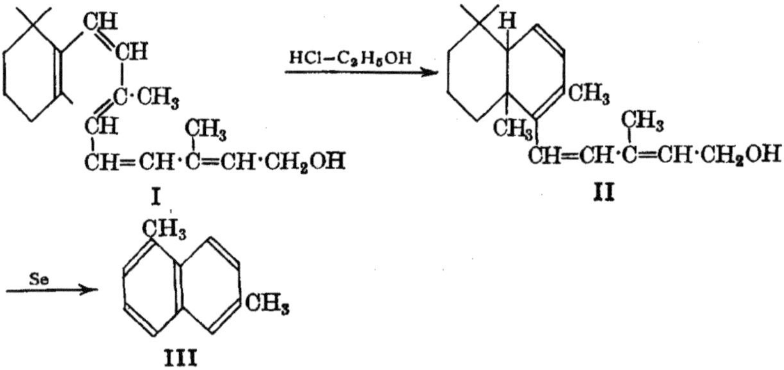

Perhydrovitamin A_1 has been synthesised from β-ionone and was shown to be identical with the compound obtained by reducing vitamin A_1.

Synthesis: Isler *et al.* (1947).

1. Methyl vinyl ketone to produce compound IV.

2. Compound V is prepared from β-ionone by means of the Darzens reaction.

Preparation of IV.

2

3. Combination of IV and V, etc.

$$CH_2\!=\!CH\cdot\overset{\overset{\displaystyle CH_3}{|}}{C}\!=\!O \xrightarrow[\text{(ii) CH}\equiv\text{CH}]{\text{(i) Na-liquid NH}_3} CH_2\!=\!CH\cdot\overset{\overset{\displaystyle CH_3}{|}}{\underset{\underset{\displaystyle \overset{+}{O}Na}{|}}{C}}\!\cdot C\!\equiv\!CH \xrightarrow{\text{H}_2\text{O}}$$

$$CH_2\!=\!CH\cdot\overset{\overset{\displaystyle CH_3}{|}}{\underset{\underset{\displaystyle OH}{|}}{C}}\!\cdot C\!\equiv\!CH \xrightarrow{\text{H}_2\text{SO}_4} CH_2OH\cdot CH\!=\!\overset{\overset{\displaystyle CH_3}{|}}{C}\!\cdot C\!\equiv\!CH$$

$$\xrightarrow{\text{C}_2\text{H}_5\text{MgBr}} BrMgOCH_2\cdot CH\!=\!\overset{\overset{\displaystyle CH_3}{|}}{C}\!\cdot C\!\equiv\!C\cdot MgBr$$

IV

Preparation of V.

$$\text{[structure]}\,CH\!=\!CH\cdot\overset{\overset{\displaystyle CH_3}{|}}{C}\!=\!O + CH_2Cl\cdot CO_2C_2H_5 \xrightarrow[\substack{\text{in liquid} \\ \text{NH}_3}]{\text{C}_2\text{H}_5\text{ONa}} \text{[structure]}\,CH\!=\!CH\cdot\overset{\overset{\displaystyle CH_3}{|}}{\underset{\underset{\displaystyle O}{\diagup\diagdown}}{C}}\!\!-\!CH\cdot CO_2C_2H_5$$

$$\xrightarrow{\text{hydrolysis}} \text{[structure]}\,CH\!=\!CH\cdot\overset{\overset{\displaystyle CH_3}{|}}{C}H\cdot CO\cdot CO_2H \xrightarrow[\substack{\text{powder under} \\ \text{red. press.}}]{\text{heat with Cu}} \left[\text{[structure]}\,CH\!=\!CH\cdot\overset{\overset{\displaystyle CH_3}{|}}{C}H\cdot CHO\right]$$

$$\xrightarrow{\text{isomerises}} \text{[structure]}\,CH_2\cdot CH\!=\!\overset{\overset{\displaystyle CH_3}{|}}{C}\!\cdot CHO$$

V

$$\underset{\text{V}}{\text{CH}_2\cdot\text{CH}=\overset{\overset{\text{CH}_3}{|}}{\text{C}}\cdot\text{CHO}} + \underset{\text{IV}}{\text{BrMgC}\equiv\text{C}\cdot\overset{\overset{\text{CH}_3}{|}}{\text{C}}=\text{CH}\cdot\text{CH}_2\text{OMgBr}}$$

$$\downarrow$$

$$\underset{\text{VI}}{\text{CH}_2\cdot\text{CH}=\overset{\overset{\text{CH}_3}{|}}{\underset{\underset{\text{OH}}{|}}{\text{C}}}\cdot\text{CH}\cdot\text{C}\equiv\text{C}\cdot\overset{\overset{\text{CH}_3}{|}}{\text{C}}=\text{CH}\cdot\text{CH}_2\text{OH}}$$

$$\downarrow \text{H}_2-\text{Pd}-\text{BaSO}_4$$

$$\underset{\text{VII}}{\text{CH}_2\cdot\text{CH}=\overset{\overset{\text{CH}_3}{|}}{\underset{\underset{\text{OH}}{|}}{\text{C}}}\cdot\text{CH}\cdot\text{CH}=\text{CH}\cdot\overset{\overset{\text{CH}_3}{|}}{\text{C}}=\text{CH}\cdot\text{CH}_2\text{OH}}$$

$$\downarrow (\text{CH}_3\cdot\text{CO})_2\text{O}$$

$$\underset{\text{VIII}}{\text{CH}_2\cdot\text{CH}=\overset{\overset{\text{CH}_3}{|}}{\underset{\underset{\text{OH}}{|}}{\text{C}}}\cdot\text{CH}\cdot\text{CH}=\text{CH}\cdot\overset{\overset{\text{CH}_3}{|}}{\text{C}}=\text{CH}\cdot\text{CH}_2\text{OCO}\cdot\text{CH}_3}$$

$$\downarrow \text{trace of I}_2 \text{ in benzene solution}$$

$$\left[\text{CH}_2\cdot\text{CH}\cdot\overset{\overset{\text{CH}_3}{|}}{\underset{\underset{\text{OH}}{}}{\text{C}}}=\text{CH}\cdot\text{CH}=\text{CH}\cdot\overset{\overset{\text{CH}_3}{|}}{\text{C}}=\text{CH}\cdot\text{CH}_2\text{O}\cdot\text{CO}\cdot\text{CH}_3 \right]$$

$$\downarrow -\text{H}_2\text{O}$$

$$\underset{\text{IX}}{\text{CH}=\text{CH}\cdot\overset{\overset{\text{CH}_3}{|}}{\text{C}}=\text{CH}\cdot\text{CH}=\text{CH}\cdot\overset{\overset{\text{CH}_3}{|}}{\text{C}}=\text{CH}\cdot\text{CH}_2\text{O}\cdot\text{CO}\cdot\text{CH}_3}$$

$$\downarrow \text{hydrolysis}$$

$$\text{CH}=\text{CH}\cdot\overset{\overset{\text{CH}_3}{|}}{\text{C}}=\text{CH}\cdot\text{CH}=\text{CH}\cdot\overset{\overset{\text{CH}_3}{|}}{\text{C}}=\text{CH}\cdot\text{CH}_2\text{OH}$$

Vitamin B$_6$

Another Nmae: Pyridoxin/Adermin

Molecule Formula:$C_8H_{11}O_3N$

- obtained from rice bran and yeast.

- it cures dermatitis in rats.

- Pyridoxin behaves as a weak base.

- Application of the Zerewitinoff method showed the presence of three active hydrogen atoms.

- When treated with **diazomethane**, pyridoxin formed a monomethyl ether which, on **acetylation**, gave a diacetyl derivative. It therefore appears that the three oxygen atoms in pyridoxin are present as hydroxyl groups, and since one is readily methylated, this one is probably phenolic. This conclusion is supported by the fact that pyridoxin gives the ferric chloride colour reaction of phenols. Thus the other two hydroxyl groups are alcoholic.

- Examination of the **ultraviolet absorption** spectrum of pyridoxin showed that it is similar to that of 3-hydroxypyridine. It was therefore inferred that pyridoxin is a pyridine derivative with the phenolic group in position 3.

- **Lead tetra-acetate** has no action on the monomethyl ether of pyridoxin, this leads to the conclusion that the two alcoholic groups are not on adjacent carbon atoms in a side-chain.

- When this methyl ether is *very carefully* **oxidised** with alkaline potassium permanganate, the product is a methoxypyridinetricarboxylic acid, $C_9H_7O_7N$. This acid gave a blood-red colour with ferrous sulphate, a reaction which is characteristic of pyridine-2-carboxylic acid; thus one of the three carboxyl groups is in the 2-position.

- When the methyl ether of pyridoxin was oxidised with alkaline permanganate under the usual conditions, the products were carbon dioxide and the anhydride of a dicarboxylic acid, $C_8H_5O_4N$; thus these two carboxyl groups are in the ortho-position.

- Furthermore, since this anhydride, on hydrolysis to its corresponding acid, did not give a red colour with ferrous sulphate, there is no carboxyl group in the 2-position. It therefore follows that, on decarboxylation, the tricarboxylic acid eliminates the 2-carboxyl group to form the anhydride; thus the tricarboxylic acid could have either of the following structures.

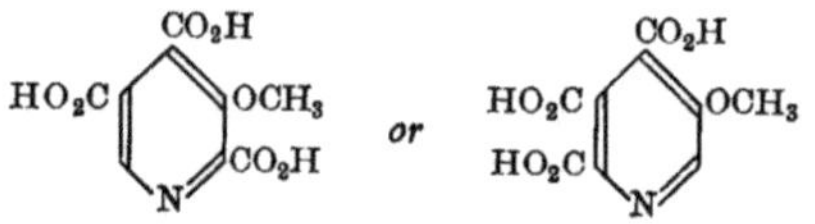

CO₂H   CO₂H

HO₂C⟨ ⟩OCH₃ *or* HO₂C⟨ ⟩OCH₃
    CO₂H   HO₂C

- Now pyridoxin methyl ether contains three oxygen atoms (one as methoxyl and the other two alcoholic) ; it is therefore possible that two carboxyl groups in the tricarboxylic acid could arise from two CH_2OH groups, and the third from a methyl group, *i.e.,* pyridoxin could be either of the following :

CH₂OH   CH₂OH

HOCH₂⟨ ⟩OH *or* HOCH₂⟨ ⟩OH
    CH₃   CH₃

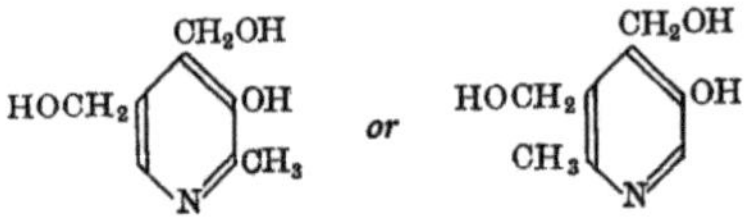

A decision between the two structures was made on the following evidence.

- When pyridoxin methyl ether was oxidised with barium permanganate, the product was a dicarboxylic acid, $C_3H_9O_5N$, which did not give a red colour with ferrous sulphate; thus there is no carboxyl group in the 2-position.

- Also, since the dicarboxylic acid formed an anhydride and gave a phthalein on fusion with resorcinol, the two carboxyl groups must be in the *ortho*position.

6

- Furthermore, analysis of both the dicarboxylic acid and its anhydride showed the presence of a methyl group. Thus the structure of this dicarboxylic acid is either I or II.

I   II

Kuhn *et al.* (1939) showed that the anhydride was that of I from its formation by the oxidation of 4-methoxy-3-methyl-isoquinoline (a synthetic compound of known structure).

Hence, on the foregoing evidence, pyridoxin is

pyridoxin

Synthesis: Harris and Folkers (1939)

ethoxyacetyl-   cyano-
acetone   acetamide

pyridoxin

Vitamins

Structure of Pyridoxal and Pyridoxamine:

Vitamins

Vitamin H

❖ **Another Name: Biotins**

- Bios, an extract of yeast, was shown to be necessary for the growth of yeast.

- It was then found that bios consisted of at least two substances, and two years later, Miller showed that three substances were present in bios.

- The first of these was named Bios **I**, and was shown to be mesoinositol.

- The second constituent, named Bios **IIA**, was then shown to be *13-alanine* or pantothenic acid.

- The third substance, named Bios IIB, was found to be identical with *biotin*, a substance that had been isolated by Kogl *et al.* as the methyl ester from egg-yolk.

- Biotin is a vitamin, being necessary for the growth of animals.

- In **1940,** du Vigneaud *et al.* isolated from liver a substance which had the same biological properties as biotin.

- Kligl *et al.* (1943) named their extract from egg-yolk a-biotin, and that from liver β-biotin. Both compounds have the same molecular formula $C_{10}H_{16}O_3N_2S$.

β-Biotin ($C_{10}H_{16}O_3N_2S$), m.p. 230-232°,

- behaves as a saturated compound.

- β-Biotin forms a monomethyl ester $C_{11}H_{18}O_3N_2S$ which, on hydrolysis, gives an acid the titration curve of which corresponds to a monocarboxylic acid; thus the formula of β-biotin may be written $C_9H_{15}ON_2S \cdot CO_2H$.

- When heated with barium hydroxide solution at 140°C, β-biotin is hydrolysed to carbon dioxide and diaminocarboxylic acid $C_9H_{18}O_2N_2S$ which, by the action of carbonyl chloride, is reconverted into β-biotin. These reactions suggest that β-biotin contains a cyclic ureide structure.

9

β-biotin   diamino-compound

- Furthermore, since the diaminocarboxylic acid condenses with phenanthraquinone to form a quinoxaline derivative, it follows that the two amino groups are in the 1:2-positions, and thus the cyclic ureide is five-membered.

- When this diaminocarboxylic acid is oxidised with alkaline permanganate, adipic acid is produced. One of the carboxyl groups in adipic acid was shown to be that originally present in β-biotin as follows.

- When the carbomethoxyl group of the methyl ester of β-biotin was replaced by an amino-group by means of the Curtius reaction (ester hydrazide –4- azide –4- urethan NH_2), and the product hydrolysed with barium hydroxide solution, a triamine was obtained which did not give adipic acid on oxidation with alkaline permanganate. It was therefore inferred that β-biotin contains a —$(CH_2)_4 \cdot CO_2H$ side-chain (n-valeric acid side-chain).

- The UV spectra of the quinoxaline derivative (formed from phenanthraquinone and the diaminocarboxylic acid) showed that it was a quinoxaline, I, and not a dihydroquinoxaline, II; thus the diaminocarboxylic could be III but not IV.

- It therefore follows that the n-valeric acid side-chain cannot be attached to a carbon atom joined to an amino-group.

- The nature of the sulphur atom in β-biotin was shown to be of the thioether type (*i.e.*, C—S—C) since:

(i) Oxidation of β-biotin with hydrogen peroxide produced a sulphone.

(ii) When the methyl ester of β-biotin was treated with methyl iodide, a sulphonium iodide was formed.

- β-biotin does not contain a double bond; hence, from its molecular formula, it was deduced that β-biotin contains two rings and 2db as as uride and carbonyl.

- When heated with Raney nickel, β-biotin formed *dethiobiotin* by elimination of the sulphur atom. Hydrolysis with hydrochloric acid, gave a diaminocarboxylic acid which, on oxidation with periodic acid, gave pimelic acid. These results can be

explained by assuming that the sulphur atom is in a five-membered ring and the n-valeric acid side-chain is in the position shown.

Further evidence for this structure is given by the fact that the exhaustive methylation of the diaminocarboxylic acid, followed by hydrolysis, gave δ-(2-thienyl)-valeric acid . the structure of this compound was confirmed by synthesis.

thiophen

glutaric anhydride

δ-(2-thienyl)-valeric acid

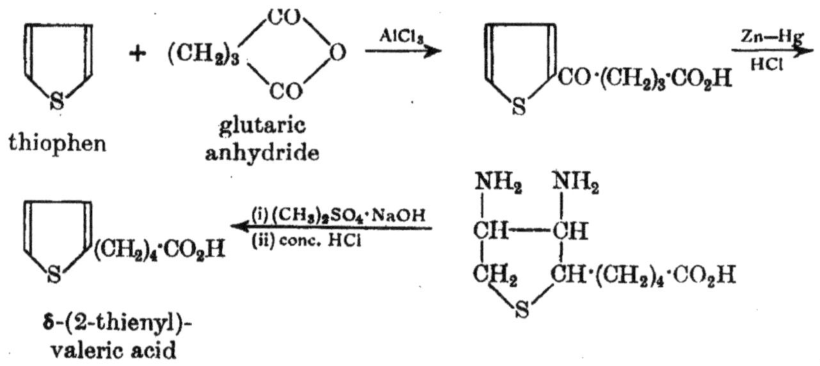

The above structure for β-biotin has been confirmed by synthesis (Harris _et al._, 1943, 1944).

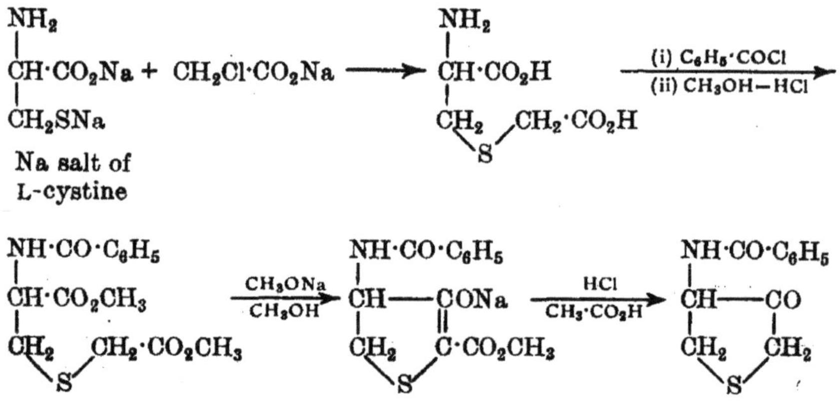

Na salt of
L-cystine

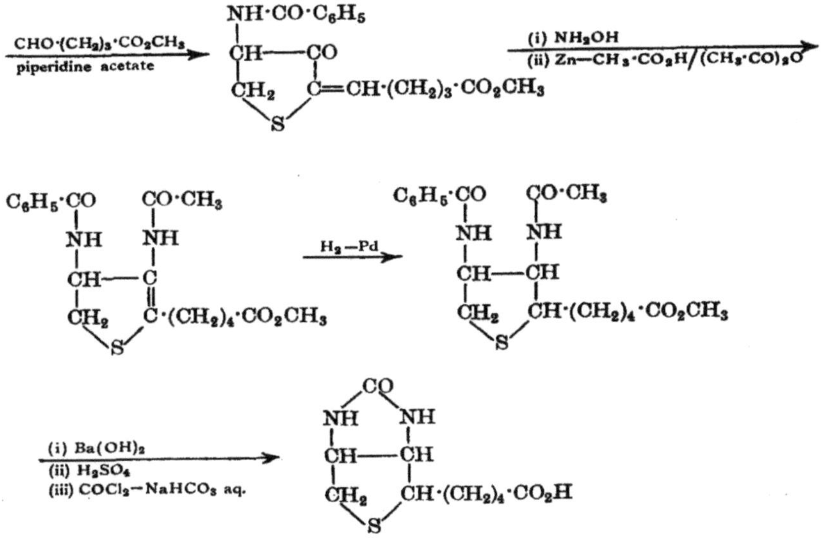

$$\xrightarrow[\text{piperidine acetate}]{\text{CHO·(CH}_2)_3\text{·CO}_2\text{CH}_3}$$

$$\xrightarrow[\text{(ii) Zn—CH}_3\text{·CO}_2\text{H/(CH}_3\text{·CO)}_2\text{O}]{\text{(i) NH}_2\text{OH}}$$

NH·CO·C$_6$H$_5$

CH——CO

CH$_2$ C=CH·(CH$_2$)$_3$·CO$_2$CH$_3$

S

C$_6$H$_5$·CO   CO·CH$_3$

NH   NH

CH——C

CH$_2$ C·(CH$_2$)$_4$·CO$_2$CH$_3$

S

$$\xrightarrow{\text{H}_2\text{—Pd}}$$

C$_6$H$_5$·CO   CO·CH$_3$

NH   NH

CH——CH

CH$_2$ CH·(CH$_2$)$_4$·CO$_2$CH$_3$

S

$$\xrightarrow[\substack{\text{(ii) H}_2\text{SO}_4 \\ \text{(iii) COCl}_2\text{—NaHCO}_3\text{ aq.}}]{\text{(i) Ba(OH)}_2}$$

CO

NH   NH

CH——CH

CH$_2$ CH·(CH$_2$)$_4$·CO$_2$H

S

Vitamin B₁

❖ **Another Name: Thiamine/Aneurin**

❖ **Molecular formula:$C_{12}H_{18}N_4OCl_2S$.**

- One member of the water-soluble vitamin B complex.

- It is the absence of thiamine which is the cause of beriberi in man. thus this vitamin is the antineuritic factor. hence the name *aneurin*.

- Rice polishings and yeast have been the usual sources of thiamine; eggs are also a rich source.

- Thiamine is obtained crystalline in the form of its salts; the chloride hydrochloride has been shown to have the molecular formula $C_{12}H_{18}N_4OCl_2S$.

- When treated with a sodium sulphite solution saturated with sulphur dioxide at room temperature, thiamine is decomposed quantitatively into two compounds which, for convenience, we shall label A and B.

$$C_{12}H_{18}ON_4Cl_2S + Na_2SO_3 \rightarrow \underset{A}{C_6H_9ONS} + \underset{B}{C_6H_9O_3N_3S} + 2NaCl$$

Compound A, C_6H_9ONS:

This compound shows basic properties,

- It does not react with nitrous acid, it was inferred that the nitrogen atom is in the tertiary state. Functional nature of the oxygen atom was shown to be alcoholic, *e.g.,* when A is treated with hydrochloric acid, a hydroxyl group is replaced by a chlorine atom.

- UV spectrum of the chloro derivative is almost the same as that of the parent (hydroxy) compound, this suggests that the hydroxyl group is in a side-chain.

- The sulphur did not give the reactions of a mercapto compound nor of a sulphide; in fact, the stability of this sulphur atom led to the suggestion that it was in a heterocyclic ring.

Vitamins

- It was confirmed by the fact that A has an UV spectrum of Compound A was characteristic of a thiazole.

- R. R. Williams *et al.* (1935) found that oxidation of A with nitric acid gives the compound $C_5H_5O_2NS$, which can also be obtained by the direct oxidation of thiamine with nitric acid.

- Williams *et al.* showed that this oxidation product was a monocarboxylic acid, and found that it was identical with 4-methylthiazole-5-carboxylic acid, I, a compound already described in the literature (WOhmann, 1890).

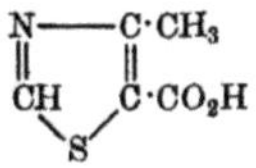

I	II

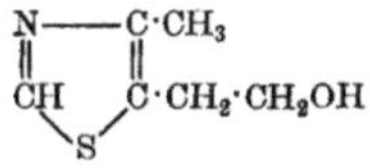

- From this it follows that A has a side-chain of two carbon atoms with alcoholic OH group. Therefore side-chain could be either —$CH_2 \cdot CH_2OH$ or —$CHOH.CH_3$ oxidized to COOH.

- The -$CHOH.CH_3$ was cancaled by the fact that A does not give the iodoform test, and that A is not optically active. Thus A was given structure **II**, and this has been confirmed by synthesis(William *et al.*).

(i)

$$\begin{array}{c} CH_3 \\ | \\ CO \\ | \\ CH^- \ Na^+ \\ | \\ CO_2C_2H_5 \end{array} + BrCH_2 \cdot CH_2OC_2H_5 \longrightarrow \begin{array}{c} CH_3 \\ | \quad CO_2C_2H_5 \\ | \quad | \\ CO \cdot CH \cdot CH_2 \cdot CH_2OC_2H_5 \end{array} \xrightarrow{SO_2Cl_2}$$

$$\begin{array}{c} CH_3 \ CO_2C_2H_5 \\ | \quad\quad | \\ CO-CCl \cdot CH_2 \cdot CH_2OC_2H_5 \end{array} \xrightarrow[\text{hydrolysis}]{\text{"ketonic}"} \begin{array}{c} CH_3 \\ | \\ CO \cdot CHCl \cdot CH_2 \cdot CH_2OC_2H_5 \end{array}$$

(ii)

$$\begin{array}{c} \overset{N|H}{\underset{\parallel}{CH}} \quad HO \vdots C \cdot CH_3 \\ \searrow \quad \quad \parallel \\ S|H \quad \quad C \cdot CH_2 \cdot CH_2OC_2H_5 \\ \quad \quad \quad Cl \end{array} \longrightarrow H_2O + HCl +$$

thioformamide

$$\begin{array}{c} N\!\!-\!\!C \cdot CH_3 \\ \parallel \quad\quad \parallel \\ CH \quad C \cdot CH_2 \cdot CH_2OC_2H_5 \\ \searrow_S\nearrow \end{array}$$

$$\xrightarrow{HCl} \begin{array}{c} N\!\!-\!\!C \cdot CH_3 \\ \parallel \quad\quad \parallel \\ CH \quad C \cdot CH_2 \cdot CH_2Cl \\ \searrow_S\nearrow \end{array} \xrightarrow[H_2O]{\text{boil with}} \begin{array}{c} N\!\!-\!\!C \cdot CH_3 \\ \parallel \quad\quad \parallel \\ CH \quad C \cdot CH_2 \cdot CH_2OH \\ \searrow_S\nearrow \\ \textbf{A} \end{array}$$

Compound B, $C_2H_9O_3N_3S$:

• This was shown to be a sulphonic acid.

• when heated with water under pressure at 200°, B gives sulphuric acid;

• it also forms sodium sulphite when heated with concentrated sodium hydroxide solution.

• On treatment with nitrous acid, B evolves nitrogen ; thus B contains one or more amino-groups.

• Analysis of the product showed that one amino-group is present in B. The evolution of nitrogen was slow, and the reaction of B with benzoyl chloride was also slow, this suggests that B contains an amidine structure NH-CH=N

• Williams *et al.* (1935) heated B with hydrochloric acid at 150° under pressure, and

$$\underset{\textbf{B}}{C_6H_9O_3N_3S} + H_2O \xrightarrow{HCl} \underset{\textbf{C}}{C_6H_8O_4N_2S} + NH_3$$

16

- Formation of ammonia indicates the replacement of an amino-group by a hydroxyl group. This type of reaction is characteristic of 2- and 6-aminopyrimidines ; it was therefore inferred that B is a pyrimidine derivative.

- Ultraviolet absorption spectrum of compound C was similar to that of synthetic 6-hydroxypyrimidines ; thus B is probably a 6-aminopyrimidine.

- When B is reduced with sodium in liquid ammonia, a sulphonic acid group is eliminated with the formation of an amino dimethyl pyrimidine.

- Comparison of the ultraviolet absorption spectrum of this product with various synthetic compounds showed that it was 6-amino-2,5-dimethylpyrimidine, and this was confirmed by synthesis (Williams *et al.*, 1937).

- Thus B is 6-amino-2,5-dimethylpyrimidine with one hydrogen atom replaced by a sulphonic acid group.

- When thiamine is treated with sodium in liquid ammonia, one of the products is the diamino derivative D, $C_6H_{10}N_4$

- Compound D was identified as 6-amino-5-aminomethyl-2-methylpyrimidine by comparison with the UV spectra of methylated aminopyrimidines of known structure.

- Thus, in compound D, there is an amino-group instead of the sulphonic acid group in B. Williams therefore concluded that the sulphonic acid group (in B) is joined to the methyl group at position 5. This was confirmed by treating 5-ethoxymethyl-6-hydroxy-2-methyl pyrimidine with sodium sulphite, whereby 6-hydroxy-2-methylpyrimidyl,5-methane

sulphonic acid was obtained, and this was shown to be identical with compound C.

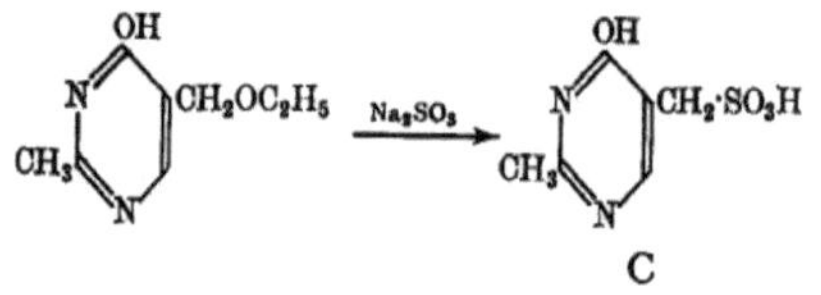

C

Thus B has the following structure:

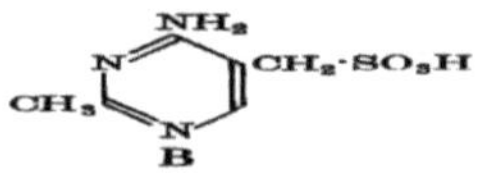

B

This structure is confirmed by synthesis (Grewe, 1936).

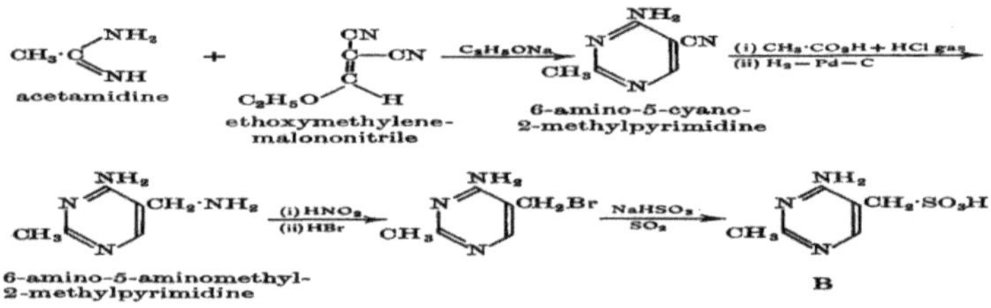

acetamidine   ethoxymethylene-malononitrile   6-amino-5-cyano-2-methylpyrimidine

6-amino-5-aminomethyl-2-methylpyrimidine   B

Final problem is: How are fragments A and B united in thiamine?

- The sulphonic acid group in B is introduced during the fission of thiamine with sodium

  sulphite; thus the point of attachment of fragment B is at the CH_2 group at position 5.

- the formation of compound D, fragment B must be linked to the nitrogen atom of fragment

  A; in this position, the nitrogen atom of the thiazole ring is in a quaternary state.

- If B been connected to A through a carbon–carbon bond, the c-c bonf atom of the latter,

  was not be easily undergo fission of this by means of sodium and liquid ammonia, nor for

  the fact that thiamine does not form a *dihydrochloride*. Thus the chloride hydrochloride of

  thiamine is

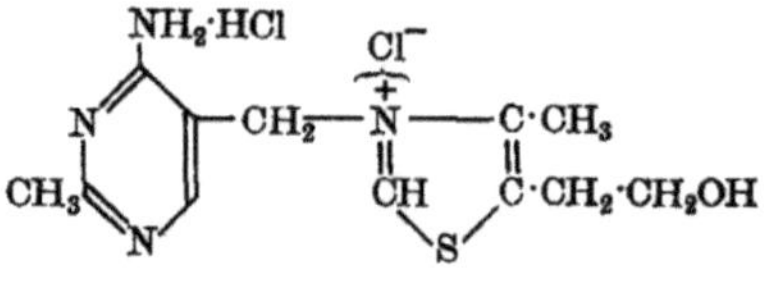

thiamine chloride hydrochloride

18

This structure has been confirmed by synthesis, *e.g.*, that of Williams *et al.*
(1936, 1937).

(i) $\begin{array}{l} CO_2C_2H_5 \\ CH_2 \cdot CH_2OC_2H_5 \end{array}$ + $H \cdot CO_2C_2H_5$ $\xrightarrow{Na}$ $\begin{array}{l} CO_2C_2H_5 \\ CH \cdot CH_2OC_2H_5 \\ CHO \end{array}$ $\xrightarrow[C_2H_5ONa]{CH_3 \cdot C \equiv NH}^{NH_2}$

(ii)

Vitamin C

❖ **Another Name: Ascorbic Acid**

Synthesis of Vitamin C

D-glucose → (+)-sorbitol

(−)-sorbose → diacetone-(−)-sorbose

2-ketogulonic-acid

L-ascorbic acid